Jacob Bigelow

Brief Expositions of Rational Medicine

To which is Prefixed the Paradise of Doctors, a Fable

Jacob Bigelow

Brief Expositions of Rational Medicine
To which is Prefixed the Paradise of Doctors, a Fable

ISBN/EAN: 9783744782098

Printed in Europe, USA, Canada, Australia, Japan

Cover: Foto ©berggeist007 / pixelio.de

More available books at **www.hansebooks.com**

BRIEF EXPOSITIONS

OF

RATIONAL MEDICINE:

TO WHICH IS PREFIXED

The Paradise of Doctors,

A FABLE.

BY

JACOB BIGELOW, M. D.,

LATE PRESIDENT OF THE MASSACHUSETTS MEDICAL SOCIETY, PHYSICIAN OF
THE MASSACHUSETTS GENERAL HOSPITAL, ETC.

SECOND EDITION.

NEW YORK:
SAMUEL S. & WILLIAM WOOD,
No. 389 BROADWAY.
1860.

PHILLIPS, SAMPSON & CO.,
In the Clerk's Office of the District Court of the District of Massachusetts

Stereotyped by
HOBART & ROBBINS,
New England Type and Stereotype Foundery,
BOSTON.

Dedication,	3
The Paradise of Doctors; A Fable,	7
Brief Expositions of Rational Medicine,	23

FELLOW OF THE ROYAL COLLEGE OF PHYSICIANS, LONDON,
PHYSICIAN OF THE QUEEN'S HOUSEHOLD, &C. &C.

My dear Sir:

The distinguished and influential position which you hold in regard to principles, many of which are advocated in this little publication, renders it proper that I should present it to your notice as an humble auxiliary in the promotion of a just, and, I hopefully trust, a growing conviction in the public mind, as to the true mission and powers of the medical art.

It is known to you that it was my intention to have published in this country an edition of your very able volume entitled "Nature and Art in the Cure of Disease," in connection with some other publications of like tendency which have appeared on this side of the Atlantic, and to have embodied the whole under the title of "Rational Medicine." Of this plan, as well as name, I had the pleasure to receive your approval and your concurrence in its execution. But after the whole was prepared, and placed in the publisher's hands here, the unforeseen appearance of a New York edition of your work rendered superfluous the proposed undertaking.

The world will duly appreciate the labor and learning which, during half a century, you have brought to the aid of true medical philosophy, and, in a particular manner, the impartial investigations which you have lately made in regard to the part performed by nature in the cure of diseases. Convictions in a measure similar to your own have, at the same time, found their way into other minds, and generally in a near proportion to the testimony afforded by prolonged experience. Twenty-three years ago I read before the Medical Society of this State a Discourse on Self-limited Diseases, which, I have reason to believe, was not without some influence at the time and since on the minds of the profession here. This discourse was afterwards incorporated, with other essays, in a volume entitled "Nature in Disease." I now hope that the crowning and convincing testimonies afforded by your noble work on the comparison of nature and art in the cure of disease will be instrumental in causing the extravagances of a so-called heroic and overactive practice on the one hand, and of a nugatory and ignorant practice on the other, to be replaced by something which may deserve the name of RATIONAL MEDICINE.

If an apology is due for so far departing from the accustomed gravity of science as to introduce at the commencement of this little book a fable called *The Paradise of Doctors*, it must be derived from the fact that, in this age of overproduction in all departments of literature, the public ear is sometimes attracted by exaggeration to give its attention afterwards to more chastened expositions of the truth.

  I am, dear Sir,
    With much respect and regard,
      Yours,

THE PARADISE OF DOCTORS.

A FABLE.

READ AT THE ANNUAL DINNER OF THE MASSACHUSETTS MEDICAL SOCIETY,
MAY 26, 1858.

IT happened, once, that a general awakening took place among the physicians, druggists and citizens, of the quiet old State of Massachusetts, during which it was discovered that a great and culpable neglect had long been prevalent throughout the community in regard to the important duty of taking physic. A conviction fell upon all that it was now imperatively necessary that every man, woman and child, should proceed at once and habitually, in sickness and in health, to take three times as much medicine as they had taken before. This new revelation, explained and enforced by competent authorities, quickened into sudden activity every department of industry connected with the preparing, prescribing and dispensing, of drugs. The repose of cities was disturbed, in a manner not before known, by the rattling of doctors' carriages and the braying of apothecaries' mortars. Messengers were seen rapidly traversing streets and roads in all directions, bearing prescriptions and compositions. Nurses' wages were doubled, and cooks were transformed into nurses. All things gave evidence that a great and portentous reform had come over the land.

In all places of business and amusement, in the street and in the drawing-room, physic was the paramount subject of conversation. Newspapers neglected to announce the arrival of steamers, and the brawls of Congress, that they might find place for the last astonishing cures, and the most newly-discovered specifics. Sympathetic intercommunications and experiences were imparted, and listened to with untiring avidity. Many luxuries unknown before found their way into society, dinners were regularly medicated, wines scientifically sophisticated, and desserts were made up of **conserves,** electuaries and **dinner-pills.** The atmosphere was **redolent with the** incense **of aloes** and myrrh.

Clergymen and moralists forgot that men were sinful; it was quite enough that they were bilious. Bile was regarded as the innate and original sin, which was **to be extirpated with fire and** physic **even from the** new-born **child. Nobody** was aware **that bile is necessary to** life; no two persons were agreed as to **what** the term *bilious* meant; it was something **insidious,** mysterious and awful. **Some held that it consisted** in having **too** much bile; others in having too little. **According** to some, the bile was held back **in** the blood; according to others, it was absorbed ready formed into the blood. **Fierce schisms and sects** were generated **on the question wh**o, and whether any, were exempt **from its** contaminating presence. The bon vivant, after his night's carouse, furnished abundant demonstrations of its existence **on** the following morning. A healthy laborer, who had had the temerity **to** boast of his freedom from bilious taint or suspicion, was con victed and brought to his senses by the ordeal of a dozen grains of tartar-emetic.

On the exchange, brokers postponed their stocks and bonds, that they might publish daily lists of the prices of drugs. Fortunes were made and lost in drug speculations. A man grew rich by a patent for manufacturing Peruvian bark out of pine saw-dust. Gilded pills, of various **weight** and potency, passed as a circulating medium, and were freely taken at the shops in payment for better goods. Finally, the physicians did not attempt to eat or sleep, but barely found time to enter their daily professional charges. They were worshipped and run after, by both sick and well, as the legitimate vehicles of medicine, and were ignominiously deserted if in any case they ventured to pronounce medicine unnecessary.

The fame of these doings went abroad, and Massachusetts acquired the enviable celebrity of **being the** Paradise of medical men. The doctors in New Hampshire, and the druggists in New **York,** hearing of the success of their professional brethren in this quarter, began to abandon their establishments and remove into Massachusetts. The example was followed in other states; new recruits were drawn from the counter and the plough, and in a short time the country and city were inundated by swarms of medical practitioners of all denominations. Agreeably to the acknowledged law of commerce and political economy, that demand and supply necessarily regulate each other, the business of many persons, which had undergone an undue exaggeration, was at length found rapidly to decline under increasing competition, and the aggregate receipts of the year were found, to the cost of not a few disciples of Esculapius, to be less than they had ever been before. Medicines became drugs, and the Paradise of Doctors became an excellent place for doctors to starve in. Nevertheless, although the market was as much glutted as the people, still a large surplus both of zeal and physic remained to be worked off in some way.

Meanwhile, the revival went on, and its effects began to tell upon the faces and movements of the people. There was a deficiency in the will to undertake, and the power to execute, even common enterprises. Men went languidly to their respective places of business, or stayed at home if it was their day to take a **purgative** or an emetic. Purses were found to be lightened, and **the** contour of persons grew sensibly less. In one thing only the economy of living was promoted: owing to the decline of appetite, the consumption of food was much diminished. Under this order of things it was noticed that labor and exercise were little in vogue, and people betook themselves in preference to the occupation of doing nothing. A small number, it is true, made **a desperate** effort to effect a change by doubling their doses of physic; but the result did not encourage a repetition of the experiment. At last a cholera came, and, although a forty-drug power was promptly brought to bear upon it, the mortality was greater than it had ever been known to be **before.**

Nevertheless, weak-minded men and strong-minded women failed not to harangue audiences in the streets on the astonishing powers of medicine. Spirit-rappers were summoned to evoke from their rest the **heroic** shades of Rush and Bouillaud, Sangrado, **Morrison and Brandreth.** These distinguished worthies exhorted their followers **not to shrink** or falter under the trials to which they **were** subjected, but **rather** to redouble their perseverance, until the truth of the faith which they held should be established by the testimony **of their martyrdom in its** cause.

At length **a meeting** accidentally took place between two old shipmasters, one **of** whom had lost overboard **his barrel of beef, and** the other his medicine-chest, in a gale of **wind at the commencement of** their passage. **On examination and comparison of** their respective **crews, the** contrast was so marked between **the ruddy faces** of the latter, **and the lantern-jaws of the former,** that a general **mutiny sprang up in both** crews against the **further tolerance of the** physic-taking **part** of their **duty.** The contagious insurrection **spread from** Fort Hill **to Copp's Hill; and on** the following night several medicine-chests **were thrown** overboard **by men in the disguise of** South-sea **Islanders.**

The spark which had struck the magazine caused the whole population to explode. A universal mass-meeting was called upon Boston Common, and protracted through several days and nights. Agitators, reformers and stump-orators, delivered their harangues, and defined their positions. Many speakers advocated an immediate application to the Legislature, calling on them to prohibit, by an especial act, all further traffic in drugs. One, more violent than the rest, demanded that the meeting should resolve itself into a committee of vigilance, for the purpose of making a descent upon the apothecaries' shops, and emptying the contents of their bottles into the streets. He was willing to allow to offenders themselves the option to quit within twenty-four hours, or swallow their own medicines. A more moderate citizen said he rose in support of the general sentiment, but would offer an amendment, that, in the contemplated destruction, an exception should be made in favor of Bourbon whiskey. A few of the advocates of the policy lately prevalent attempted to make themselves heard; but their voices were so attenuated, by

THE PARADISE OF DOCTORS. 15

the long use of jalap and salts, that they failed to produce any considerable impression.

An old lady, whose shrill voice drew immediate attention, protested against violent measures of all kinds, and moved, as a middle course, that resort should be had to homœopathy. It never did any harm, and was very comforting, especially when well recommended by the physician. It cured her child of the measles in six weeks, and herself of a broken leg in six months, during which time she had two hundred and ninety-five visits, and took more than fifteen hundred globules. She had walked to the meeting on her crutches to exhibit to the assembly the astonishing powers of the Hahnemannic system. Here she was interrupted by a bluff marketer, who somewhat rudely pronounced homœopathy to be a great humbug, since, but a short time before, his child had eaten part of a raw pumpkin, and was seized with convulsions; and the physician who was sent for, instead of taking measures to dislodge the offending cause, took out a little book, and remarking to the by-standers that "like cures like," proceeded to prescribe the hundred millionth part of another pumpkin. — The next person who rose was a manufacturer, who had calculated that the homœopathic profit on the cost of the raw material was altogether unreasonable. He had himself expended seventy-five dollars in a quarter of a grain of belladonna, so divided as to keep off scarlet fever; but found, after all, that he had not bought enough, for his children had the disease a little worse than any of their neighbors.

At last an old gentleman, moderately endowed with common sense, got up, and inquired if there was no such thing in the world as *rational medicine*, and whether nothing could be made acceptable to the public but extremes of absurdity. He asked if it was necessary that every theologian should be a Calvinist or an atheist, or every voter at the polls an abolitionist or a fire-eater. He had had the good fortune to know several very sensible, straight-forward physicians, who gave medicine where it was necessary, and omitted to give it where it was unnecessary or detrimental. He deprecated the routine practice which, without understanding the nature of a disease, or the necessities of the existing case, inflicted a daily or hourly dose of medicine, sometimes actual and sometimes nominal, but always at the cost of the patient. Medicine, in its place, was a good thing, but proved a bad thing when we got too much of it. He had himself had the misfortune to be several times sick, and, during the continuance of his disease, felt much more gratified on those days in which it was announced that he was to take no medicine, than when tartar emetic was replaced by calomel, and calomel by colchicum, aconite, and the last new remedy. If patients and their friends were ignorant and unreasonable, it might sometimes be necessary to deal with a fool according to his folly; but he believed that sensible men and women were gratified by being regarded and treated as reasonable beings. It was a mistake in medical men to suppose that their influence or social position could be improved by the mystery which they observed, and the activity with which they harassed their patients. In Great Britain, an island where the people subsist largely on blue pills and black draughts, the doctors were never known to attain the high aristocratic rank which **was** occasionally accorded to successful bankers, jurists **and** generals. On the contrary, the country was overflowed with starved apothecaries and physicians advertising for situations as travelling servants. He thought one of the greatest misapplications of human industry was in the production of superfluous drugs **and drug-**dispensers. **He did** not believe in the transmutation of metals, **but was a** great believer in their transportation **In the form of** calomel, the **city of New** Orleans alone **had** swallowed **up some** hundred tons of the **quicksilver of** Spain and South America. Palaces **were being built in** various **cities alike from poisonous arsenic** and harmless **sarsaparilla.** A century hence **the mines** of gold will be **sought for, not** in California, but in the **cemeteries of the old** cities, where **it has been** geologically deposited under the industry of **dentists.**

He believed that the experienced and intelligent part of the **medical** profession had long since arrived at the conclusion that many diseases were self-limited, and **that** time and nature had quite as much to do as art in the process of their cure. Skilful physicians were always wanted to inform the sick of the character of their diseases, and of the best mode of getting through them; and their skill consisted not in the abundance of their nominal remedies, but in the judgment with which a few remedies were administered or withheld, and in the general safe conduct of the patient. Some diseases are curable by art, and others are not; yet, in the treatment of all diseases, there is a right method and a wrong, and too much activity is quite as injurious as too little. A good shipmaster or pilot could often navigate his vessel in safety, though he could not cure the storm by which its safety was endangered. He believed that medicine would have fulfilled its true mission when doctors should have enlightened the public on the important fact that there are certain things which medicine can do, and certain other things which it cannot do, instead of assuming for it the power to do impossibilities. Among the good effects which must ensue from this diffusion of light would be the disappearance of quackery from the world; for quackery consists almost wholly in medication. And the more physicians lend themselves to formal, superfluous and mysterious drugging, the more nearly do they approach to being quacks themselves. He considered physicians an important and necessary class, to whose charge the sick always had been, and always would be, committed. He would gladly cleanse the profession from the fanaticism of heroic doctors on the one hand, and of moonstruck doctors on the other, and would replace these forms of delusion by a discriminating, sincere, intelligent and rational, course of treating diseases.

The old gentleman sat down, and his speech seemed good in the eyes of his audience. Resolutions were moved and adopted to the effect that it was unbecoming a free and enlightened people to be drug-ridden or globule-ridden, and recommending recourse to temperance, exercise, regularity and rational medicine, whenever it happened that medical treatment was necessary.

The meeting quietly dissolved, and its members returned to their respective homes, most of them satisfied that the revival was past, and that medicine was not altogether the one thing needful. In a short time the price of drugs fell in the market, while that of provisions advanced. The New Hampshire doctors and the New York druggists, finding their occupations gone, returned to the places from which they respectively came. The surplus of indigenous medical men went off to California, or retired to cultivate the earth in the interior counties. Faces assumed a more vigorous and healthy aspect, and the country once more resounded with the music of the axe and the hammer, and the cheerful rattling of knives and forks. Steam-engines, which had been erected for the pulverization of drugs, were attached to saw-mills and spinning-jennies. Last of all, a noble and useful art, which had long been depressed under the effects of its own exaggeration, was enabled once more to raise its respectable head, and to regain the confidence of society, under the name of RATIONAL MEDICINE.

BRIEF EXPOSITIONS

OF

RATIONAL MEDICINE

The tendency to ultraism, which influences public opinion in great social questions, particularly of politics and theology, has been also prevalent in the affairs of practical medicine. No age has been exempt from diversity of opinion among physicians on the speculative subjects of their art; and the present period appears to be more marked than preceding ones by the opposite methods of treatment pursued by medical men in the management of disease. These methods consist, for the most part, in a vehement, officious and over-drugging system on the one hand, and an inert, evasive and nugatory practice on the other. Between these extremes the intermediate truth meets with less consideration than **it ought to** receive from unbiased and enlightened inquirers.

Extreme doctrines in practical subjects often arise from the self-interest of those who originate and first promote them. But the vehemence and fanaticism with **which they are** afterwards pursued are as often owing to the creation **of false issues,** which **divert** public attention from the **substance to the shadow, and** mystify the general **question** with **minor, partial, and** frequently **irrelevant considerations.**

The introduction **into** the English language, for example, **of the** term "allopathy," and its adoption **by some** medical writers, has had the **effect to mislead** superficial readers in regard to **the true issue of** questions connected with the treatment of disease. This word was designed by its zealous, **but** weak and incompetent, inventor to express the employment of remedies which produce phenomena *different* from those produced **by the** disease treated; whereas the term homœopathy indicated a mode of treating diseases by employing medicines which are sup**posed to** produce effects *similar* to those which it is desired to remove. This theoretical and absurd generalization, wholly unsupported **by** general facts on either side, so far as the cure of **diseases is** concerned, has acquired currency among the less enlightened part of the public, so that, at the present day, many persons consider homœopathy a sort of general law, to which **allopathy is** an exception. **And,** strange to tell, many otherwise sensible physicians have assumed the cloak thus offered to them, without perceiving that the **propriety of so doing is the** same as if the whole protestant world were to **style themselves** heretics, because the Church **of** Rome, **in** former ages, saw fit to apply to them that appellation. Allopathy is, in fact, a worthless term, which either means nothing real, or else embodies so many dissimilar and discordant elements that it serves no purpose as a descriptive or distinctive name. The occasion still exists for terms which **may** definitively express the dogmas of modern practice.

Anatomy, physiology, and to a certain extent pathology, may be considered, so far as our discoveries have advanced, to be entitled to rank with the exact sciences. But therapeutics, or the art of treating diseases, like ethics and political economy, is still a conjectural study, incapable of demonstration in many of its great processes, and subject to various and even opposite opinions in regard to the laws and means which govern its results.

The methods which, at the present day, are most prevalent in civilized countries, in the treatment of disease, may be denominated the following:

1. The *Artificial* method, which, when carried to excess, is commonly termed heroic, and which consists in reliance on artificial remedies, usually of an active character, in the expectation that they will of themselves remove diseases.

2. The *Expectant* method. This consists simply in non-interference, leaving the chance of recovery to the powers of nature, uninfluenced by interpositions of art.

3. The *Homœopathic* method. This is a counterfeit of the last, and consists in leaving the case to nature, while the patient is amused with nominal and nugatory remedies.

4. The *Exclusive* method, which applies one remedy to all diseases, or to a majority of diseases. **This** head includes hydropathy, also the **use of** various mineral waters, electrical establishments, **etc.** Drugs newly introduced, **and** especially secret medicines, frequently boast this universality of application.

5. The *Rational* **method.** This recognizes nature as the great agent in the cure of diseases, and employs art as **an** auxiliary, to be resorted to when **useful or** necessary, **and** avoided when prejudicial.

The foregoing methods, with the exception perhaps of the last, have had their trial in various **periods** and countries, and have given rise to discussions and controversies which are not terminated at the present day. The subject is too complicated **to obtain** from inquirers, out of the profession, **the amount of** attention requisite for understanding its merits; while, among medical men, consistency to pledged opinions, defects of knowledge, and considerations **of** interest, not unfrequently warp the judgment of otherwise honest and discerning persons. It is certain, moreover, that medical opinion on the treatment of disease changes much between the time of one generation and another. Any person who will take the trouble to inspect the medical journals published thirty or forty years ago will find many things, then laid down as medical truths, which are now generally admitted to be medical errors. The length of a common professional life is sufficient to disabuse most physicians of many convictions which they had received on trust, and once considered unchangeable. Yet, it does not always happen that error is replaced by truth, and it is fortunate if the delusions of ill-balanced minds are not succeeded by newer and greater delusions.

It is, nevertheless, right that intelligent and reasonable physicians should receive the confidence of the community, since they are, or should be, more qualified than other persons to undertake the care and conduct of the sick. And even if it had happened that their power was limited to merely understanding the nature of the existing disease, and the import and probable tendency of symptoms which occur from day to day, without any attempt at curative interference, still their attendance would **be** solicited **to throw** light on the grave questions of pain, sickness, and recovery, and still more of life and death. The public, however, expect something more of physicians than the power of distinguishing diseases, and of predicting their issue. They look to them for the relief **of their** sufferings, and the **cure** or removal **of** their complaints. And the vulgar estimate **of** the powers of medicine is founded on the common acceptation of the name, that medicine is the **art** of **curing** diseases. That this is a false definition, is evident from the fact that many diseases are incurable, and that one such disease must at last happen to every living man.* **A** far more just definition would **be, that** medicine **is** the art **of** understanding diseases, and **of** curing or relieving them **when possible. If** this definition were accepted, **and its truth** generally understood **by** the profession and **the** public, a **weight of** super-

* See the author's "Nature **in Disease**," page 64.

fluous responsibility on one side, and of dissatisfaction on the other, would be lifted from the shoulders of both. **It** is because physicians allow themselves to profess and vaunt more power over disease than belongs to them, that their occasional short-comings **are** made a ground of reproach with the community, and of contention among themselves.

It is now generally admitted by intelligent physicians that certain diseases, the number of which is not very great, are at once curable by medical means. Yet, there is probably no curative agent, applied to such diseases, the efficacy and even safety of which has not been warmly contested by sectarian practitioners. It is also beginning to be admitted **in this** country that certain diseases are *self-limited,** incurable now

---

* This term was first introduced **by** the writer in a discourse in 1835, already alluded **to,** with the following definition : " A self-limited disease **is one which** receives limits from **its own** nature, and not from **foreign** influences ; one which, after it has obtained foothold in the system, cannot, **in** the present state of our knowl**edge,** be eradicated **or abridged by art, but** to which there is due a certain succession of processes, to be completed in a certain time

by art, yet susceptible of recovery under natural processes, both with and without the interference of art. To this class belong a great portion of the diseases usually called acute, and likewise some, the character of which is decidedly chronic. Lastly, a vast tribe of incurable diseases takes precedence in the lists of mortality, and holds, in some form, its final sentence over the heads of all mankind. Yet, so reluctant are physicians to acknowledge these universal truths, or to admit their own incompetency, that incurable and unmanageable diseases have been complacently called *opprobria medicinæ*, as if they were exceptions to a general rule.

The great objects which medical practice professes to effect, and which there can be no doubt that it frequently does effect, are the following: 1. The cure of certain diseases. 2. The relief or palliation of all diseases. 3. The safe conduct of

which time and processes may vary with the constitution and condition of the patient, and may tend to death or to recovery, but are not known to be shortened, or greatly changed, by medical treatment."

the sick. In all these objects it sometimes fails; yet, instances of its success are sufficiently numerous to establish the necessity of the existence of medicine as a profession.

No one doubts that morbid affections, occasioned by the presence of an offending or irritating cause, are often speedily cured by the discharge or removal of that cause. And here drugs are among the principal agents which we employ. Again, no one doubts that many of the diseases of civilized life, brought on by luxury, intemperance, sedentary and intellectual labor, unhealthy residence, occupation, etc., are often wholly or partially cured by change of life, including habits, and perhaps residence. And here drugs are, for the most part, of little avail. So that it may happen that the chance of cure shall depend upon the judgment with which active **drugs are** administered, on the one hand, and **avoided or superseded,** on the other.

The **palliation of** diseases is another great practical end **of** medical science, and really occupies a large portion **of the time** and efforts of **every** medical man. When it is considered that most diseases last for days, and some of them for years, and that a large portion of mankind eventually die of some chronic or lingering disease, it will **readily be** seen that the palliation of suffering is not only called for, but really constitutes one of the most important, as well as beneficent objects of medical practice. The use **of anodynes** and anæsthetics, the obviation of various painful and distressing symptoms, the removal of annoyances, the just regulation of diet, of exertion and repose, of indulgence and restriction, the direction of moral agencies, which make up so large a part both of suffering and relief, may well afford employment to the most earnest and philanthropic physician, and obtain from the public a just appreciation of the value of his services.

The safe conduct of **the sick,** as will be seen from the last head, consists **much** more in cautionary guidance than in active interference. In the management of sickness, **the rein** is needed to direct quite as much as the spur to **excite.** People sometimes suffer from **neglect, but** more frequently from ill-judged and meddlesome attention. One of the most cogent necessities of a sick man is to be saved from the excessive and officious good will of his friends. The kindest impulses and the most benevolent intentions are liable to show themselves in ill-timed visits, fatiguing conversations, and injudicious advice or action. Intelligent and discreet physicians are sometimes driven by the importunity of friends to the adoption of active measures, or at least the semblance of them, which their own judgment informs them would be better omitted. And the case is still worse when the impulsive temperament of the physician himself, or the influence of his early education, or the dominant fashion of the place in which he resides, is so exacting in regard to activity of treatment as to make him believe that he cannot *commit* too many inflictions upon the sick, provided that, in the end, he shall be satisfied that he has *omitted* nothing.

The foregoing desiderata, the cure, the relief, and the safe conduct of patients, involve the great objects for which medicine has been striving for thousands of years. Yet, even in the present advanced state of science, physicians are not agreed as to the means by which any one of them is to be accomplished or attempted; and a man who falls sick at home or abroad is liable to get **heroic** treatment or nominal treatment, random treatment or no treatment at all, according to the hands into which he may happen to fall. It is, therefore, **desirable** that **physicians** themselves, and the public also, should obtain a satisfactory **understanding of the various diversities** of **practice which have** been already mentioned as occupying the greatest share of attention at the present day.

1. THE ARTIFICIAL METHOD. — This **mode** of treatment **is** founded on the assumption that disease can be removed by artificial means. From the earliest ages a belief has prevailed that all human maladies **are** amenable **to** control from some form of purely medical treatment; and although the precise form **has not** yet been found, so far as most diseases are concerned, yet, at this day, it continues to be as laboriously and hopefully pursued as was the elixir **vitæ** in the **middle** ages. Within the present century, books of practice gravely laid down " the indications of cure,"

as if they were things within the grasp of every practitioner. It was only necessary to subdue the inflammation, to expel the morbific matter, to regulate the secretions, to improve the nutrition, and to restore the strength, and the business was at once accomplished. What nature refused, or was inadequate to do, was expected to be achieved by the more prompt and vigorous interposition of art. The destructive tendencies of disease, and the supposed proneness to deterioration of nature herself, were opposed by copious and exhausting depletion, followed by the shadowy array of alteratives, deobstruents and tonics. Confinement by disease, which might have terminated in a few days, was protracted to weeks and months, because the importance of the case, as it was thought, required that the patient should be artificially " taken down," and then artificially " built up."

When carried to its " heroic " extent, artificial medicine undermined the strength, elicited new morbid manifestations, and left more disease than it took away. The question raised was not how much the patient had profited under his active treatment, but how much more of the same he could bear. Large doses of violent and deleterious drugs were given as long as the patient evinced a "tolerance" of them, that is, did not sink under them. The results of such cases, if favorable, like the escapes of desperate surgery, were chronicled as professional triumphs, while the press was silent on the disastrous results subsequently incurred in like cases by deluded imitators. If diseases proved fatal, or even if they were not jugulated or cut short at the outset, the misfortune was attributed to the circumstance of the remedies not being sufficiently active, or of the physician not being called in season. So great at one time, and that not long ago, was the ascendency of heroic teachers and writers, that few medical men had the courage to incur the responsibility of omitting the active modes of treatment which were deemed indispensable to the safety of the patient. This timidity on the score of omission has now, in a great measure, passed away, yet is still promoted in most cities by some heroic doctors, and still more by interested specialists, who inflict severe discipline, and levy immense contributions, on credulous persons, who are suitably alarmed at denunciations which involve the loss of sight, of hearing, **or** even of beauty.

A considerable amount of violent practice is still maintained by **routine** physicians, who, **without** going deeply into the true nature or exigencies of the case before them, assume the general ground that nothing is dangerous but neglect. Edge-tools are brought into use as if they could never be anything more than harmless playthings. **It** is thought allowable to harass the patient with daily and opposite prescriptions; to try, to abandon, to reënforce, or to reverse; to blow hot and cold on successive days; but never to let the patient alone, nor to intrust his case to the quiet guidance of nature. Consulting physicians frequently and painfully witness the gratuitous suffering, the continued nausea, the prostration of strength, the prevention of appetite, the stupefaction of the senses, and the wearisome days and nights, which **would never have** occurred had there been no such thing as officious medication. What practitioner has not seen infants screaming under the pangs of hunger, or of stimulants remorselessly applied to their tender skins, and whose only permitted chance of relief was in the continued routine of unnecessary calomel and ipecacuanha?

There is one great exception in favor of artificial and even heroic practice, well known and fully demonstrable in the art of surgery. Many defects, injuries, and diseases of the body, are, unquestionably, cured by surgical processes, which never could have got well without them. And the skilful and humane surgeon has more frequent opportunities to reflect with satisfaction on the immediate and positive results of his art than the most able physician. Yet even this satisfaction can only be measured by the fidelity with which he has performed his duty, and the conscientiousness with which he has avoided useless and hopeless operations. Happily the experience and statistical results of the best modern surgeons have had the effect to diminish greatly the amount of gratuitous suffering which was imposed by their predecessors on the unhappy subjects of their art. We see much less than was formerly seen of the cruel but unavailing operations of fanciful and interested surgery; the infliction of pain without corresponding good, the useless extirpation of malignant growths, the mutilation of miserable bodies already doomed by tuberculous and other irrecoverable conditions; deeds which have converted hospitals into inquisitions, and left the Bastile and the Hotel Dieu to contend for the palm of supremacy in the production of human suffering.

2. THE EXPECTANT METHOD. — This method, when fully carried out, admits no medication nor interference of art, but waits on time, and commits the chance of recovery to the restorative power of nature alone. The expectant practice has not been without its advocates, and volumes have been published in its favor, at different times, chiefly on the continent of Europe. That there is some basis for the doctrine of expectation is made apparent by the spontaneous recovery of animals and savages, of careless, obstinate and incredulous persons in civilized life, and of those who, in consequence of their isolated or otherwise unfavorable position, are

unable to procure "medical aid," or who, if they do procure it, obtain only that which is inoperative or absolutely detrimental. I sincerely believe that the unbiased opinion of most medical men of sound judgment and long experience is made up, that the amount of death **and** disaster in the world would be **less,** if all disease were left to itself, than it **now is under** the multiform, **reckless** and contradictory modes of practice, good and bad, **with which** practitioners of adverse denominations carry **on their** differences at **the** expense of their patients. But **there** is no probability that expectant medicine will **ever** prevail in its character as such. The amount of positive good which, in fifty centuries, art has brought to the assistance of medicine, although far more limited than we **could** desire, **is,** nevertheless, **both sufficient and worthy to** employ the talents **of the best and most** enlightened physicians.

3. THE HOMŒOPATHIC METHOD.—Homœopathy may be defined as a specious **mode of doing** nothing. While it waits on the natural progress of disease and the restorative tendency of nature

on the one hand, or the injurious advance of disease on the other, it supplies the craving for activity, on the part of the patient and his friends, by the formal and regular administration of nominal medicine. Although homœopathy will, at some future time, be classed with historical delusions, yet its tendency has undoubtedly been to undermine the reliance on heroic practice which prevailed in former times both in this country and in Europe. There was, perhaps, needed a popular delusion to institute the experiment on a sufficiently large scale to show that the sick may recover without the use of troublesome and severe medication. There are not wanting in history similar instances of good results flowing from questionable sources. The French Revolution has eventually bettered the social condition of the French people; and the Mormons have brought the wilderness of the Salt Lake to a state of productive cultivation. Yet no judicious person vindicates the doctrines of those who were prime movers in these innovations, or holds them up as worthy examples for imitation. Sir Kenelm Digby produced a beneficial reform in English surgery, and was able to banish the prevalent mode of dressing incised wounds with painful applications, by speciously going **from the** effect to the cause, and applying the active medicament, not to the wound, but to **the** weapon that did the mischief; thus giving to the former a chance to heal by the first **intention.**

There is great reason **to believe that, at the** present day, homœopathic faith is not always kept up in **its** original purity by its professors. Traces of the occasional use of very heroic **remedies** are **often** detected among the most unsuspected of its practitioners. **And it** must not be concealed that there are instances in which the temptation **is very great, even for** the most resolute convert, to come to the aid **of** the sick with reasonable and efficient doses **of real** medicine. The man must be somewhat **of a stoic who** can look upon a case **of severe colic, or** of the multiform distresses **which result** from overtasked organs of digestion, **and** quiet his **conscience with** administering inappreciable globules, instead **of** remedies.

4. THE EXCLUSIVE METHOD. — **This,** like the

heroic system, is various in its means of treatment, but differs from it in the professed universality of its peculiar applications. Hydropathy applies one remedy, cold water, to all cases. Yet, like homœopathy, it combines with its special agent a strict course of life, including exercise, temperance, regular hours, and a diet in the main simple and wholesome, though somewhat fanciful in its exclusions. The same was done so far as was proper in the previous practice of all judicious physicians. The use of cold bathing is not new, having been employed as a hygienic process from time immemorial by the civilized world. As a therapeutic agent, cold affusion was resorted to more than half a century ago, and has been practiced ever since in a greater or less degree. But the peculiar mode of applying water by packing appears to be original with Priessnitz, an ignorant German, to whom it owes its popularity. Like the Russian bath, in which alternate approaches to scalding and freezing are said to be followed at last by very delightful sensations, the hydropathic discipline, in those who have soundness of constitution sufficient to insure a healthy reaction, is followed by agreeable and often salubrious results. Yet the ineffective character of hydropathy is seen in the multitude of disappointed invalids who return unrelieved from its establishments. I have been told, by persons who have resided at Graefenberg, that funerals at that place were of constant occurrence; and it is well known that Priessnitz, himself a robust peasant, died in the prime of life, in the midst of his own water-cure.

The greatest benefit at hydropathic establishments is obtained by those who reform their mode of life by submitting to the restraints of the place. The luxurious, the indolent, the sedentary, and the erratic, improve most under a return to regular, natural, active and temperate habits. Accordingly it is found that gout, dyspepsia, lost appetite, hysteria, and the various forms of nervous irritability, furnish the most hopeful subjects for such institutions. The same patients might, in many cases, obtain the same relief in another place, by pursuing the water cure without the water.

The universality of hydropathic application has been somewhat diminished by prolonged experience. Priessnitz himself, although ignorant of science, and incapable of distinguishing one disease from another, at last became cautious in his selections, and nominally excluded diseases of the lungs from his institution.

It is not necessary to dwell upon the various exclusive modes of practice, more or less universal in their application, with which the columns of newspapers are daily filled. Mineral waters, taken at the fountain, are often of great use to those who require a journey or a change of scene. Particular springs also appear to exert a beneficial effect on particular maladies, though not panaceas for all ills. Watering-places, which combine amusement with exercise, are the temporary safety-valves of over-taxed physicians, and happily afford arks of refuge to multitudes of chronic valetudinarians. Electricity supports one or more establishments in all large cities, both in its simple form, and combined with all other imponderable agencies of mind and matter.

Few persons go uncured of chronic maladies without having given it a sufficient and **satisfactory** trial. Finally, the host of **empirical remedies, which fill** the attention of a very consider**able portion of** this quack-ridden world, leave no human maladies out of the catalogue of subjects to their mysterious power. The **drug** aloes, in its hundred **pill combinations, levies** incessant contributions **on those who** purchase the **privi**lege of being slaves to its use. Opium, variously disguised with aromatics **to** conceal its presence, gives temporary but fallacious respite **to** fatal diseases, under the deceptive names of pectorals and pulmonics.

It is superfluous to prolong the consideration **of** general and exclusive remedies. No person accustomed to witness the various morbid conditions which invade and **occupy** the human frame, active and passive, **partial** and general, trivial and dangerous, can ever consider them proper subjects **for** the same kind of treatment, unless, with Dr. Rush and Dr. **Brandreth, he** happens to be a believer **in the** unity of disease.

5. THE RATIONAL METHOD. — If no alternative were left to the physician and patient but the extreme and frequently irrational methods which have now been briefly described, practical medicine might well take its rank as a pseudo-science by the side of astrology and spiritualism. But the labors of earnest and philanthropic men, **during many centuries, though often speculative,** misguided, and terminating in error, have nevertheless elicited enough of general truth to serve **as** the foundation for a stable superstructure. And, that such truth may hereafter go on to accumulate, it must be simply and honestly sought, even when its developments do not at once promote the apparent interest of physicians, nor flatter their professional pride of opinion.

It is to sincere and intelligent observers, and not to audacious charlatans, that we are to look as the ultimate lawgivers of medical science. Our present defect is not that we know too little, but that we profess too much. We regard it as a sort of humiliation to acknowledge that we cannot always cure diseases, forgetting that in **many other sciences** mankind have made no greater advances than ourselves, and are still upon the threshold of their respective structures. Medical assumption may well feel humbled by the most insignificant diseases of the human body. Take, for example, a common furunculus or boil. No physician can, by any internal treatment, produce it where it does not exist. No physician can, by any science, explain it, and say why it **came on one limb and not upon** another. No physician can, **by** any art, cure it after it has arrived **at a certain** height. No physician can, by any art, **delay or** retain it after it has passed the climax assigned to it by nature. And what is true in regard to a boil is equally true of common pneumonia, of typhoid fever, of acute rheumatism, of cholera, and many other diseases.

In the **present state of our knowledge the** truth appears to be simply this: Certain diseases, of which the number is not very great, are curable, or have **their cure** promoted, by drugs, and by appliances which are strictly medicinal. Certain other diseases, perhaps more numerous, are curable in like manner by means which are strictly regiminal, and consist in changes of place, occupation, diet, and habits of life. Another **class** of diseases are self-limited, and can neither be **expelled from the** body by artificial means, nor retained in the body after their natural period of duration has expired. Finally, a large class of diseases have **prov**ed incurable from the beginning **of** history to the present time, and **under** some one of these the most **favored members of** the human race must finally succumb; for even curable diseases become incurable when they have **reached** a certain stage, extent, or complication.

It will be seen that the divisions last mentioned **cannot be strictly** reduced under the nomenclature **of nosologies; for** cases, and groups of cases, may **begin in one** category and end in another.

It is the part of rational medicine to study intelligently the nature, degree and tendency, of each **existing case, an**d afterwards to act, or to forbear acting, as the exigencies of such case may require. **To do** all this wisely and efficiently, the practitioner must possess, first, sufficient know ledge to diagnosticate the disease; and, secondly, sufficient sense as well as knowledge to make up a correct judgment on the course to be pursued. In the first of these, if properly educated and **experienced, he** will be **able to** make an approximation to the truth sufficient for practical purposes. In the second **he** will have to depend mainly on his well-ordered **and logical powers of** self-direction; **for he will find in the recorded** evidence of **his** predecessors quite as much to mislead as to guide him rightly. He will find many existing **cases, in which for a** time he will know not what to do, and in which his safest course will be not to do he knows not what. It **is better to resort to a little** expectancy than to rush into blind and reckless action. Nature, when not encumbered with **overwhelming burdens,** and when **not abused by** unnatural and pernicious excesses, **is, after all,** " the kindest mother still." **Art may sometimes** remove those burdens, and regulate those excesses; but it is not by imposing new burdens and instituting new excesses that **an** end so **desirable is to** be attained. Before commencing any contemplated course of treatment, in a given case, two questions should always be asked: 1. Will it do good? 2. Will it do harm? A right answer to these questions will not fail to produce a right practice.

It is the part of rational medicine to alleviate the sufferings of the sick. And for this end alone, were there no other, physicians would be necessary as a profession. For this end alone, any person knowingly about to encounter the confinement of a self-limited fever, or the lingering decay of a cancer or consumption, would invoke the guidance of a medical man whose judgment and skill were better than his own. The power of the medical art to palliate diseases is shown in a multitude of ways, active, cautious and expectant. The pain of acute pleurisy is relieved by venesection; that of pleurodynia, by anodynes and external applications. The pain of acute rheumatism is postponed by opium; that of gout, by colchicum. Synovitis is favorably affected by rest; chronic rheumatism more frequently by exercise. Demulcents, opiates, and even astringents, have their use in various irritations of the mucous membranes. Cathartics, laxatives, emetics, leeches, counter-irritants, cupping, hot and cold applications, etc., are of benefit in various local and general maladies. Yet these remedies, especially the more energetic of them, are often employed when not necessary, and become, by their degree and frequency, rather sources of annoyance than of relief. Violent cathartics are followed by increased constipation, when milder laxatives or enemata would not have induced that evil. Blisters, antimonial ointments, salivation, etc., may continue to afflict the patient long after the disease is gone. The effects of powerful depletion are felt for months, and sometimes for years. Excessive stimulation by vinous liquids may create or renew disease, or give rise to pernicious artificial wants. To prescribe blindly for **symptoms, irrespectively of their cause,** is often in the highest degree injudicious. The alvine discharges of dysentery and typhoid fever are the natural ventings of an inflamed, perhaps **ulcerated**, membrane; the pain and the **excess** may be abated by the gentlest anodynes, but the attempt to check them altogether would be like the drying up of an external ulcer, of equal dimensions, by the sudden application of astringents. The object might be attained for a day, but the result would be pernicious. Having already touched upon this subject, I have only to add, that if many of the troublesome appliances and severe exactions of modern practice were superseded by gentler, more soothing, and more natural means, a good would be done to the human race comparable to the conversion of swords into ploughshares.

It is the part of rational medicine still to strive and study for the cure of diseases; not to assume fallaciously as practical truth what has never been shown to be true, but rather to search and labor for new truth, for which it is never too late to hope. The rational physician will ever be ready to weigh and examine, candidly and carefully, new practical questions and proposed modes of treatment, whether introduced for the alleviation or the removal of diseases; and he will recollect that although nineteen out of every twenty of the new methods proposed may be worthless, yet the twentieth may **perhaps** possess some valuable quality. It is known that the most established laws of science cease to be such when **their** exceptions have been detected and **made out.** Some of the most important advances in human knowledge have been among the latest in date. **The** great American discovery of artificial anæsthesia **has been wished and** waited for by mankind ever since the flood; yet the effectual conquest **of pain is** as it were a thing of yesterday.

It **is** the part of rational medicine to require evidence for what it admits and believes. The cumbrous fabric now called therapeutic science is, in a **great** measure, built up on the imperfect testimony of credulous, hasty, prejudiced, or incompetent witnesses, such **as** have afforded authorities for books like Murray's Apparatus Medicaminum, and Hahnemann's Organon. The enormous polypharmacy of modern times is an excrescence on science, unsupported by any evidence of necessity or fitness, and of which **the** more complicated formulas are so arbitrary and useless, that, **if** by any chance they should be forgotten, not one in a hundred of them would **ever** be reinvented. And as to the **chronicles** of cure of diseases that are not yet **known to be curable,** they are written, not in the pages of philosophic **observers,** but in the tomes of compilers, the aspirations of journalists, and the columns of advertisers.

It is the part of rational medicine to enlighten the public **and** the profession in regard to the true powers of the healing art. The community require to be undeceived and reëducated, so far as to know what is true **and** trustworthy from what is gratuitous, unfounded and fallacious. And the profession themselves will proceed with confidence, self-approval and success, in proportion as they shall have informed mankind on these important subjects. The exaggerated impressions now prevalent in the world, in regard to the powers of medicine, serve only to keep the profession and the public in a false position, to encourage imposture, to augment the number of candidates **struggling** for employment, **to** burden and disappoint the community already overtaxed, to lower the standard of professional character, and **raise** empirics to the level of honest and enlightened physicians.

I AM not willing to **leave** the subject of Rational Medicine without more earnestly calling the attention of the profession in the United States to the admirable work of Sir John Forbes, already alluded **to, on Nature and Art** in the Cure of Disease, of **which an** American edition has recently appeared. No testimony of mine can be needed to make known the claims of one of the most accomplished medical scholars of Europe, **of whose** philosophic mind, vigorous perceptions, and clear, discriminating and impartial judgment, this volume is a legitimate product and a convincing evidence.

Neither is it necessary to **inform the** public that, as one of the pioneers in **the** reform now in progress, the same author published, in 1846, in the British **and** Foreign Medical **Review,** an elaborate article, bearing the **title of** "Young Physic," which, though somewhat startling in the novelty of its positions, and by many disapproved for the credit given to statements of questionable parties, had, nevertheless, the effect, both in England and this country, to increase public attention to inquiries which the present volume is so well adapted to satisfy. Some of the concluding propositions of this able article, not everywhere accessible to American readers, are reprinted as an Appendix to the present little volume.

In France, perhaps more than any other country, the natural history of disease has been studied irrespectively of artificial influences. When the modes of rigorous investigation, to which, in that country, the laws of morbid affections have been submitted, shall have been carried as far as possible into the more complex and difficult subject of therapeutics, they will at least help to guard it from the errors of premature and speculative generalization to which the medical world has in all ages been prone.*

* The numerical method, so advantageously applied by Louis and his successors to determine approximately the pathological character of diseases, cannot be well applied to the more complex

Of the contributions to rational medicine which have been made in this country, it gives me gratification to refer to various able, just and eloquent discourses and essays, by my friends and others, which have appeared at different

subject of medical treatment, without great caution and reserve on the score of inference. Things submitted to this ordeal must be such as have some obvious connection with each other; otherwise we are liable to be led into error by the most careful observation and analysis. Such would be the case if we were to attempt to settle numerically the questions, whether medicine applied to a weapon promoted the cure of a wound, — whether the hanging of witches had favorably affected the duration of epidemics, — whether it is safe for sailors to go to sea on Friday, etc. These are questions which are settled by the common sense of an enlightened age, and not by numerical analysis. The extensive numerical trials, made in different countries, on the effect of bleeding in pneumonia, have not yet afforded results fully satisfactory to inquirers. It is worthy of notice that questions of relief are more promptly settled than questions of duration and of safety. A man suffering the orthopnœa of pleurisy will lie down and breathe with comparative ease after venesection; and this is sufficient motive for the reasonable use of that remedy. But the length and safety of the disease, under different modes of treatment afford questions yet to be settled, if at all, by a vast amount of observation, by competent persons, under the various differences of constitution, degree, complication, season, age, etc.

times. Among these should be distinguished a Discourse " On the Condition, Prospects and Duties, of the Medical Profession," by Edward Reynolds, M. D., published in 1841; a Discourse entitled "Search Nature and Know her Secrets," by Augustus A. Gould, M. D., 1855; a Discourse on " Nature in Disease," by Benjamin E. Cotting, M. D., 1852; a Prize Essay* on " Rational Therapeutics," by Professor Worthington Hooker, of New Haven, 1857; also a previous volume, by the same able and intelligent writer, entitled " Physician and Patient," 1849.

If it be permitted to refer to an extra-professional authority, bearing, nevertheless, the marks of much intelligent observation, I would cite, in behalf of the same cause, " The Rational Doctor," in the Household Words of Charles Dickens.

* The prize thus awarded was liberally instituted by Dr. B. E. Cotting, one of the writers above named.

APPENDIX.

The following are the principal remarks of Sir J. Forbes, appended as a sort of recapitulation to his article already referred to in the *British and Foreign Medical Review*, vol. xxi., p. 262. They are there submitted by him as things to be reflected and acted on by the medical profession. A great portion of them, though perhaps not all, are applicable in this country as well as in England.

"1. To endeavor to ascertain, much more precisely than has been done hitherto, the natural course and event of diseases, when uninterrupted by artificial interference; in other words, to attempt to establish a true natural history of human diseases.

"2. To reconsider and study afresh the physiological and curative effects of all our therapeutic agents, with a view to obtain more positive results than we now possess.

"3. To endeavor to establish, as far as is practicable, what diseases are curable, and what are not; what are capable of receiving benefit from medical treatment, and what are not; what treatment is the best, the safest, the most agreeable; when it is proper to administer medicine, and when to refrain from administering it, etc., etc.

"4. To endeavor to introduce a more philosophical and accurate view of the relations of remedies to the animal economy and to diseases, so as to dissociate in the minds of practitioners the notions of *post hoc* and *propter hoc.**

"The general adoption by practitioners, in recording their experience, of the system known by the name of the *Numerical Method*, is essential to the attainment of the ends proposed in the preceding paragraphs, as well as in many that are to follow.

"5. To endeavor to banish from the treatment of acute and dangerous diseases, at least, the ancient axiom, *melius anceps remedium quam nullum,*† and to substitute in its place the safer and wiser dogma,

---

* Subsequent and consequent.

† A doubtful remedy is better than none.

that, where we are not certain of an indication, we should give nature the best chance of doing the work herself, by leaving her operations undisturbed by those of art.

"6. To endeavor to substitute, for the monstrous system of Polypharmacy now universally prevalent, one that is, at least, vastly more simple, more intelligible, more agreeable, and, it may be hoped, one more rational, more scientific, more certain, and more beneficial.

"7. To direct redoubled attention to hygiene, public and private, with the view of preventing diseases on the large scale, and, individually, in our sphere of practice. Here the surest and most glorious triumphs of medical science are achieving and to be achieved.

"8. To inculcate generally a milder and less energetic mode of practice, both in acute and chronic diseases; to encourage the Expectant preferably to the Heroic system, — at least where the indications of treatment are not manifest.

"9. To discountenance all active and powerful medication in the acute exanthemata and fevers of specific type, as small-pox, measles, scarlatina, typhus, etc., until we obtain some evidence that the course of these diseases can be beneficially modified by remedies.

"10. To discountenance, as much as possible, and **eschew** the habitual use — without any sufficient reason — of certain powerful medicines, in large doses, in a multitude of different diseases; **a practice now** generally prevalent and fraught with **the** most baneful consequences.

"This is one **of** the besetting sins **of** English practice, **and** originates **partly** in false theory, and partly in the desire to see manifest and strong effects resulting from the action of medicines. Mercury, iodine, colchicum, antimony, also purgatives in general, and blood-letting, are frightfully misused in this manner.

"11. To encourage **the** administration of simple, feeble, or altogether powerless, non-perturbing medicines, in all **cases in which** drugs are prescribed *pro forma*, for **the** satisfaction of the patient's mind, and not with the view of producing any direct **remedial** effect.

"**One would** hardly think such a caution necessary, were it not **that every-day** observation proves it to be so. The system **of giving and also** of *taking* drugs capable **of** producing some obvious effect —

on the sensations, at least, **if not on the** functions — has become so inveterate **in this country, that even our** *placebos* **have, in** the hands **of our modern doctors, lost their** original quality **of** harmlessness, **and** often **please their** very patients **more by** being made unpleasant!

"12. To make every effort, not **merely to destroy the** prevalent system **of giving a vast quantity and** variety of unnecessary **and useless drugs, — to say** the least of them, — **but to encourage extreme simplicity in the** prescription of medicines **that seem to** be requisite.

"Our system **is here greatly and radically wrong.** Our officinal formulæ **are already** most absurdly **and** mischievously **complex, and our fashion is to double and** redouble **the** existing complexities. **This system is a most serious** impediment **in the** way **of** ascertaining the precise **and** peculiar **powers (if any) of the** individual **drugs, and thus** interferes, in the most important **manner, with the progress of** therapeutics.

"We are **aware of the** arguments that are adduced in defence **of medicinal** combinations. **We do not deny that** some **of these combinations** are beneficial, **and, therefore, proper ; but there cannot be a** question as to the enormous evils, speaking generally, resulting from them. Nothing has a greater tendency to dissociate practical medicine from science, and to stamp it as a *trade*, than this system of pharmaceutical artifice. It takes some years of the student's life to learn the very things which are to block up his path to future knowledge. A very elegant prescriber is seldom a good physician.

"13. To endeavor to break through the routine habit, universally prevalent, of prescribing certain determinate remedies for certain determinate diseases, or symptoms of diseases, merely because the prescriber has been taught to do so, and on no better grounds than conventional tradition.

"Even when the medicines so prescribed are innocuous, the routine proceeding impedes real knowledge by satisfying the mind, and thus producing inaction. When the drugs are potent, the crime of mischief is superadded to the folly of empiricism. In illustration, we need merely notice the usual reference, in this country, of almost all chronic diseases accompanied with derangement of the intestinal functions, to 'affection of the liver,' and the consequent prescription of *mercury* in some of its forms. We do not hesitate to say that this theory is as far wrong as the practice founded on it is injurious; we can hardly further enhance the amount of its divarication from the truth.

"14. To place in a more prominent point of view the great value and importance of what may be termed the physiological, hygienic, or natural system of curing diseases, especially chronic diseases, in contradistinction to the pharmaceutical or empirical drug-plan generally prevalent. This system, founded as it is on a more comprehensive inquiry into *all* the remote and exciting causes of disease, and on a more thorough appreciation of *all* the discoverable disorders existing in all the organs and functions of the body, does not, of course, exclude the use of drugs, but regards them (generally speaking) as subservient to hygienic, regimenal, and external means, such as the rigid regulation of the diet, the temperature and purity of the air, clothing, the mental and bodily exercise, etc., baths, friction, change of air, travelling, change of occupation, etc., etc.

"15. To endeavor to introduce a more comprehensive and philosophical system of Nosology, at least in chronic diseases, whereby the practitioner may be led less to consider the name of a disease, or some one symptom, or some one local affection in a disease, than the disease itself; that is, *the whole of the derangements* existing in the body, and which it is his object to remove, if possible.

"16. To teach teachers to teach the rising generation of medical men that it is infinitely more *practical* to be master of the elements of medical science, and to know diseases thoroughly, than to know by rote a farrago of receipts, or to be aware that certain doctors, of old or of recent times, have said that certain medicines are good for certain diseases.

"17. Also to teach students that no systematic or theoretical classification of diseases, or of therapeutic agents, ever yet promulgated, is true, or anything like the truth, and that none can be adopted as a safe guide in practice. It is, however, well that these systems should be known, as most of them involve some pathological truths, and have left some practical good behind them.

"18. To endeavor to enlighten the public as to the actual powers of medicines, with a view to reconciling them to simpler and milder plans of treatment. To teach them the great importance of having their diseases treated in their earliest stages, in order to obtain a speedy and efficient cure; and, by some modification in the relations between the patient and practitioner, to encourage and facilitate this early application for relief.

"19. To endeavor to abolish the system of medical practitioners being paid by the amount of medicine sent in to their patients; and even the practice of keeping and preparing medicines in their own houses.

"Were a proper system introduced for securing a good education to chemists and druggists, and for examining and licensing them, — all of easy adoption, — there could be no necessity for continuing even the latter practice; while the former is one so degrading to the medical character, and so frightfully injurious to medicine in a thousand ways, that it ought to be abolished forthwith, utterly, and forever.

"20. Lastly, and above all, to bring up the medical mind to the standard necessary for studying, comprehending, appreciating, and exercising, the most complex and difficult of the arts that are based on a scientific foundation — the art of Practical Medicine. And this can only be done by elevating the preliminary and fundamental education of the Medical Practitioner."

www.ingramcontent.com/pod-product-compliance
Lightning Source LLC
Chambersburg PA
CBHW031609110426
42742CB00037B/1466